Consumer's Guide to Radon Reduction
How to Fix Your Home

OVERVIEW

Reduce Radon Levels in Your Home

Radon is the leading cause of lung cancer for non-smokers and the second leading cause of lung cancer for the general population. The Surgeon General and EPA recommend testing for radon and reducing radon in homes that have high levels. Fix your home if your radon level is confirmed to be 4 picocuries per liter, pCi/L, or higher. Radon levels less than 4 pCi/L still pose a risk, and in many cases may be reduced. If you smoke and your home has high radon levels, your risk of lung cancer is especially high.

Select a State Certified and/or Qualified Radon Mitigation Contractor

Choose a qualified radon mitigation contractor to fix your home. Start by checking with your state radon office. Many states require radon professionals to be licensed, certified, or registered. You also can contact private radon proficiency programs for lists of privately certified radon professionals in your area. See pages 4 and 17 for more information.

Radon Reduction Techniques Work

Radon reduction systems work. Some radon reduction systems can reduce radon levels in your home by up to 99 percent. Most homes can be fixed for about the same cost as other common home repairs. Your costs may vary depending on the size and design of your home and which radon reduction methods are needed. Get an estimate from one or more qualified radon mitigation contractors. Hundreds of thousands of people have reduced radon levels in their homes.

Maintain Your Radon Reduction System

Maintaining your radon reduction system takes little effort and keeps the system working properly and radon levels low. See page 13 for more information.

INTRODUCTION

T his booklet is for people who tested their home for radon and have elevated radon levels — 4 pCi/L or higher. This booklet can help you:

- Select a qualified radon mitigation contractor to reduce the radon levels in your home.

- Determine an appropriate radon reduction method.

- Maintain your radon reduction system.

Your state radon office can provide information on how to test your home or how to locate a qualified radon professional https://www.epa.gov/radon/find-information-about-local-radon-zones-and-state-contact-information. EPA's A Citizen's Guide to Radon and The Home Buyer's and Seller's Guide to Radon have information on radon testing. Both documents are available at https://www.epa.gov/radon/publications-about-radon.

HOW RADON ENTERS YOUR HOME

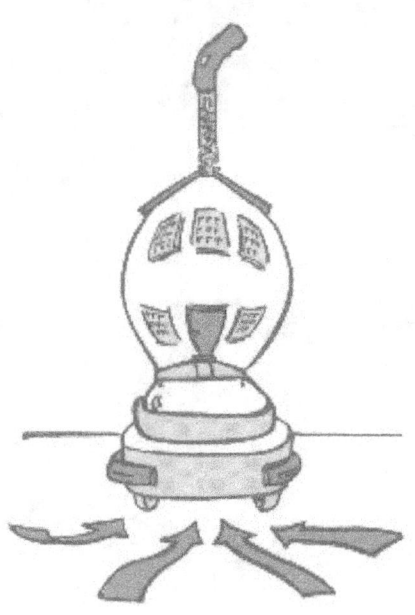

Radon is a naturally occurring radioactive gas produced by the breakdown of uranium in soil, rock, and water. Air pressure inside your home is usually lower than pressure in the soil around your home's foundation. Because of this difference in pressure, your home acts like a vacuum, drawing radon in through foundation cracks and other openings.

Radon also may be present in well water and can be released into the air in your home when water is used for showering and other household uses. In most cases, radon entering the home through water is a small risk compared with radon entering your home from the soil. In a small number of homes, the building materials — such as granite and certain concrete products — can give off radon, although building materials rarely cause radon problems by themselves. In the United States, radon gas in soils is the principal source of elevated radon levels in homes.

RADON IS A CANCER-CAUSING, RADIOACTIVE GAS

Radon is estimated to cause tens of thousands of lung cancer deaths each year. **In fact, the Surgeon General has warned that radon is the second leading cause of lung cancer in the United States.** Only smoking causes more lung cancer deaths. If you smoke and your home has high radon levels, your risk of lung cancer is especially high.

WHAT DO YOUR RADON TEST RESULTS MEAN?

Any radon exposure has some risk of causing lung cancer. The lower the radon level in your home, the lower your family's risk of lung cancer. The amount of radon in the air is measured in pCi/L.

The U.S. Congress has set a long-term goal that indoor radon levels be no more than outdoor levels; about 0.4 pCi/L of radon is normally found in the outside air. EPA recommends fixing your home if the results of one long-term test or the average of two short-term tests show radon levels of 4 pCi/L or higher. With today's technology, radon levels in most homes can be reduced to 2 pCi/L or below. You also may want to consider fixing if the level is between 2 and 4 pCi/L.

A short-term test remains in your home for two days to 90 days, whereas a long-term test remains in your home for more than 90 days. All radon tests should be taken for a minimum of 48 hours. A short-term test will yield faster results, but a long-term test will give you a better understanding of your home's year-round average radon level.

EPA recommends two categories of radon testing. One category is for concerned homeowners or occupants whose home is not for sale; refer to EPA's pamphlet "A Citizen's Guide to Radon" for testing guidance. The second category is for real estate transactions; refer to EPA's pamphlet "Home Buyer's and Seller's Guide to Radon," which provides guidance and answers to some common questions. Both documents are available at https://www.epa.gov/radon/publications-about-radon.

SELECTING A RADON TEST KIT

Since you cannot see or smell radon, special equipment is needed to detect it. When you're ready to test your home, contact your state radon office for information on locating qualified test kits or qualified radon testers. You also can order test kits and obtain information at http://sosradon.org/ There are two types of radon testing devices. Passive radon testing devices do not need power to function. These include charcoal canisters, alpha-track detectors, charcoal liquid scintillation devices, and electret ion chamber detectors. Both short- and long-term passive devices are generally inexpensive. Active radon testing devices require power to function and usually provide hourly readings and an average result for the test period. These include continuous radon monitors and continuous working level monitors, and these tests may cost more. A state or local official can explain the differences between the devices and recommend ones that are most appropriate for your needs and expected testing conditions. Make sure to use a radon testing device from a qualified laboratory.

WHY HIRE A CONTRACTOR?

EPA recommends that you have a qualified radon mitigation contractor fix your home because lowering high radon levels requires specific technical knowledge and special skills. Without the proper equipment or technical knowledge, you could actually increase your radon level or create other potential hazards and additional costs. However, if you decide to do the work yourself, get information on appropriate training courses from your state radon office.

WILL ANY CONTRACTOR DO?

EPA recommends that you use a certified or qualified radon mitigation contractor trained to fix radon problems. You can determine a service provider's qualifications to perform radon measurements or to mitigate your home in several ways. First, check with your state radon office. Many states require radon professionals to be licensed, certified or registered, and to install radon mitigation systems that meet state requirements. Most states can provide you with a list of knowledgeable radon service providers doing business in the state (https://www.epa.gov/radon/find-information-about-local-radon-zones-and-state-contact-information).

In states that don't regulate radon services, ask the contractor if they hold a professional proficiency or certification credential, and if they follow industry consensus standards, such as the American Society for Testing and Materials, ASTM, Standard Practice for Installing Radon Mitigation Systems in Existing Low-Rise Residential Buildings, E2121. You can contact private proficiency programs for lists of privately certified professionals in your area. Such programs usually provide members with a photo ID card, which indicates their qualifications and the ID card's expiration date. For more information on private proficiency programs, visit https://www.epa.gov/radon/find-radon-test-kit-or-measurement-and-mitigation-professional#what, or contact your state radon office.

HOW TO SELECT A CONTRACTOR

Get Estimates
Choose a contractor to fix a radon problem just as you would choose someone to do other home repairs. It is wise to get more than one estimate, to ask for references, and to contact some of those references to ask if they are satisfied with the contractor's work. Also, ask your state radon office or your county or state consumer protection office for information about the contractors.

Use this checklist when evaluating and comparing contractors and ask the following questions:

☐ ☐ **Will the contractor provide references or photographs, as well as test results of before and after radon levels of past radon reduction work?**

☐ ☐ **Can the contractor explain what the work will involve, how long it will take to complete, and exactly how the radon reduction system will work?**

☐ ☐ **Does the contractor charge a fee for any diagnostic tests? Although many contractors give free estimates, they may charge for diagnostic tests. These tests help determine what type of radon reduction system should be used and in some cases are necessary, especially if the contractor is unfamiliar with the type of house structure or the anticipated degree of difficulty. See "Radon Reduction Techniques" on page 8 for more on diagnostic tests.**

☐ ☐ **Did the contractor inspect your home's structure before giving you an estimate?**

☐ ☐ **Did the contractor review the quality of your radon measurement results and determine if appropriate testing procedures were followed?**

Compare the contractor's proposed costs and consider what you get for your money, taking into account: a less expensive system may cost more to operate and maintain; a less expensive system may have less aesthetic appeal; a more expensive system may be best for your home; and, the quality of the building material will affect how long the system lasts.

Do the contractors' proposals and estimates include:

YES NO

☐ ☐ **Proof of state certification, professional proficiency or certification credentials?**

☐ ☐ **Proof of liability insurance and being bonded, and having all necessary licenses to satisfy local requirements?**

☐ ☐ **Diagnostic testing prior to design and installation of a radon reduction system?**

☐ ☐ **Installation of a warning device to caution you if the radon reduction system is not working correctly?**

☐ ☐ **Testing after installation to make sure the radon reduction system works well?**

☐ ☐ **A guarantee to reduce radon levels to 4 pCi/L or below? And if so, for how long?**

The Contract

Ask the contractor to prepare a contract before any work starts. Read the contract before you sign it. Make sure everything in the contract matches the original proposal. The contract should describe exactly what work will be done prior to

and during the installation of the system, what the system consists of, and how the system will operate. Many contractors provide a guarantee that they will adjust or modify the system to reach a negotiated radon level (e.g., 2 pCi/L or less). Carefully read the conditions of the contract describing the guarantee. Consider optional additions to your contract that may add to the initial cost of the system, but may be worth the extra expense. Typical options might include an extended warranty, a service plan or improved aesthetics.

Important information that should appear in the contract includes:

☐ **The total cost of the job, including all taxes and permit fees; how much, if any, is required for a deposit; and when payment is due in full.**

☐ **The time needed to complete the work.**

☐ **An agreement by the contractor to obtain necessary permits and follow required building codes.**

☐ **A statement that the contractor carries liability insurance and is bonded and insured to protect you in case of injury to persons, or damage to property, while the work is being done.**

☐ **A guarantee that the contractor will be responsible for damage during the job and cleanup after the job.**

☐ **Details of any guarantee to reduce radon below a negotiated level.**

☐ **Details of warranties or other optional features associated with the hardware components of the mitigation system.**

☐ **A declaration stating whether any warranties or guarantees are transferable if you sell your home.**

☐ **A description of what the contractor expects the homeowner to do, such as make the work area accessible, before work begins.**

WHAT TO LOOK FOR IN A RADON REDUCTION SYSTEM

In selecting a radon reduction method for your home, you and your contractor should consider several things, including: how high your initial radon level is, the costs of installation and system operation, your home size, and your foundation type.

Installation and Operating Costs

Most types of radon reduction systems cause some loss of heated or air conditioned air, which could increase your utility bills. How much your utility bills increase will depend on the climate you live in, what kind of reduction system you select, and how your home is built. Systems that use fans are more effective in reducing radon levels; however, they will slightly increase your electric bill.

RADON REDUCTION TECHNIQUES

There are several methods a contractor can use to lower radon levels in your home. Some techniques prevent radon from entering your home while others reduce radon levels after it has entered. EPA generally recommends methods that prevent the entry of radon. **Soil suction,** for example, prevents radon from entering your home by drawing the radon from below the home and venting it through a pipe, or pipes, to the air above the home where it is quickly diluted.

Any information that you may have about the construction of your home could help your contractor choose the best system. Your contractor will perform a visual inspection of your home and design a system that considers specific features of your home. If this inspection fails to provide enough information, the contractor may need to perform **diagnostic tests** during the initial phase of the installation to help develop the best radon reduction system for your home. For instance, your contractor can use chemical smoke to find the source and direction of air movement. A contractor can learn air flow sources and directions by watching a small amount of smoke that he or she shot into holes, drains, sumps or along cracks. The sources of air flow show possible radon routes. A contractor may have concerns about backdrafting of combustion appliances when considering radon mitigation options, and may recommend that the homeowner have the appliances checked by a qualified inspector.

Another type of diagnostic test is a soil communication test. This test uses a vacuum cleaner and chemical smoke to determine how easily air can move from one point to another under the foundation. By inserting a vacuum cleaner hose in one small hole and using chemical smoke in a second small hole, a contractor can see if the smoke is pulled down into the second hole by the force of the vacuum cleaner's suction. Watching the smoke during a soil communication test helps a contractor decide if certain radon reduction systems would work well in your home.

Whether diagnostic tests are needed is decided by details specific to your home, such as the foundation design, what kind of material is under your home, and by the contractor's experience with similar homes and similar radon test results.

Home Foundation Types

Your home type will affect the kind of radon reduction system that will work best. Homes are generally categorized according to their foundation design. For example: **basement; slab-on-grade**, concrete poured at ground level; or **crawlspace,** a shallow unfinished space under the first floor. Some homes have more than one foundation design feature. For instance, it is common to have a basement under part of the home and to have a slab-on-grade or crawlspace under the rest of the home. In these situations, a combination of radon reduction techniques may be needed to reduce radon levels to below 4 pCi/L.

BASEMENT   SLAB-ON-GRADE   CRAWLSPACE

Radon reduction systems can be grouped by home foundation design. Find your type of foundation design above and read about which radon reduction systems may be best for your home.

Basement and Slab-on-Grade Homes

In homes that have a basement or a slab-on-grade foundation, radon is usually reduced by one of four types of soil suction: **subslab suction, drain-tile suction, sump-hole suction,** or **block-wall suction.**

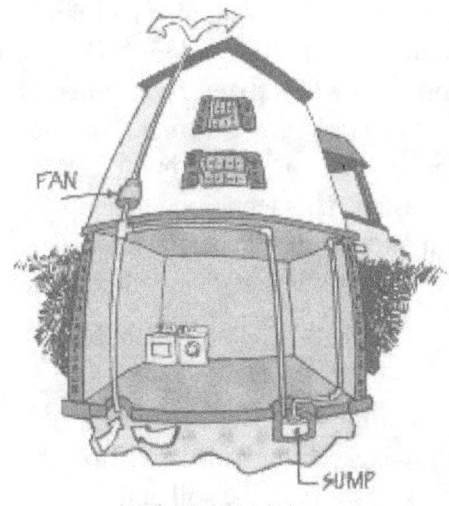

FAN

SUMP

SUBSLAB SUCTION

Active subslab suction — also called **subslab depressurization** — is the most common and usually the most reliable radon reduction method. One or more suction pipes are inserted through the floor slab into the crushed rock or soil underneath. They also may be inserted below the concrete slab from outside the home. The number and location of suction pipes that are needed depends on how easily air can move in the crushed rock or soil under the slab and on the strength of the radon source. Often, only a single suction point is needed.

A contractor usually gets this information from visual inspection, from diagnostic tests or from experience. A radon vent fan connected to the suction pipes draws the radon gas from below the home and releases it into the outdoor air while simultaneously creating a negative pressure or vacuum beneath the slab. Common fan locations include unconditioned home and garage spaces, including attics and the exterior of the home.

Passive subslab suction is the same as active subslab suction except it relies on natural pressure differentials and air currents instead of a fan to draw radon up from below the home. Passive subslab suction is usually associated with radon-resistant features installed in newly constructed homes. **Passive subslab suction** is generally not as effective in reducing high radon levels as active subslab suction.

Some homes have **drain tiles or perforated pipe** to direct water away from the foundation of the home. Suction on these tiles or pipes is often effective in reducing radon levels.

One variation of subslab and drain tile suction is **sump-hole suction**. Often, when a home with a basement has a sump pump to remove unwanted water, the sump can be capped so that it can continue to drain water and serve as the location for a radon suction pipe.

Block-wall suction can be used in basement homes with hollow block foundation walls. This method removes radon and depressurizes the block wall, similar to subslab suction. This method is often used in combination with subslab suction.

Crawlspace Homes

An effective method to reduce radon levels in crawlspace homes involves covering the earth floor with a high-density plastic sheet. A vent pipe and fan are used to draw the radon from under the sheet and vent it to the outdoors. This form of soil suction is called **submembrane suction, and when properly applied is the most effective way to reduce radon levels in crawlspace homes.** Another less-favorable option is active crawlspace depressurization, which involves drawing air directly from the crawlspace using a fan. This technique generally does not work as well as submembrane suction and requires special attention to combustion appliance backdrafting and sealing the crawlspace from other portions of the home. It also may result in increased energy costs due to loss of conditioned air from the home.

In some cases, radon levels can be lowered by ventilating the crawlspace passively, or actively, with the use of a fan. Crawlspace ventilation may lower indoor radon levels both by reducing the home's suction on the soil and by diluting the radon beneath the home. Passive ventilation in a crawlspace is achieved by opening vents or installing additional vents. Active ventilation uses a fan to blow air through the crawlspace instead of relying on natural air circulation. In colder climates, for either passive or active crawlspace ventilation, water pipes, sewer lines and appliances in the crawlspace may need to be insulated against the cold. These ventilation options could result in increased energy costs for the home.

Other Types of Radon Reduction Methods

Other radon reduction techniques that can be used in any type of home include: sealing, house or room pressurization, heat recovery ventilation and natural ventilation.

Sealing cracks and other openings in the foundation is a basic part of most approaches to radon reduction. Sealing the cracks limits the flow of radon into your home, thereby making other radon reduction techniques more effective and cost-efficient. It also reduces the loss of conditioned air. EPA does not recommend the use of sealing alone to reduce radon because, by itself, sealing has not been shown to lower radon levels significantly or consistently. It is difficult to identify and permanently seal the places where radon is entering. Normal settling of your home opens new entry routes and reopens old ones.

House or room pressurization uses a fan to blow air into the basement, or living area from either upstairs or outdoors. It attempts to create enough pressure at the lowest level indoors — in a basement, for example — to prevent radon from entering into the home. The effectiveness of this technique is limited by home construction, climate, other appliances in the home and occupant lifestyle. In order to maintain enough pressure to keep radon out, the doors and windows at the lowest level must not be left opened, except for normal entry and exit. This approach generally results in more outdoor air being introduced into the home, which can cause moisture intrusion and energy penalties. Consequently, this technique should only be considered after the other, more-common techniques have not sufficiently reduced radon.

A **heat recovery ventilator,** or **HRV**, also called an **air-to-air heat exchanger,** can be installed to increase ventilation, which will help reduce the radon levels in your home. An HRV will increase ventilation by introducing outdoor air while using the heated or cooled air being exhausted to warm or cool the incoming air. HRVs can be designed to ventilate all or part of your home, although they are more effective in reducing radon levels when used to ventilate only the basement. If properly balanced and maintained, they ensure a constant degree of ventilation throughout the year. HRVs also can improve air quality in homes that have other indoor pollutants. There could be significant increase in the heating and cooling costs with an HRV, but not as great as ventilation without heat recovery.

Some **natural ventilation** occurs in all homes. By opening windows, doors and vents on the lower floors, you increase the ventilation in your home. This increase in ventilation mixes outdoor air with the indoor air containing radon, and can result in reduced radon levels. However, once windows, doors and vents are closed, radon concentrations most often return to previous values within about 12 hours. Natural ventilation in any type of home should normally be regarded as only a temporary radon reduction approach because of the following disadvantages: loss of conditioned air and related discomfort; greatly increased costs of conditioning additional outside air; and security concerns.

CHECKING YOUR CONTRACTOR'S WORK

Below is a list of basic installation requirements that your contractor should meet when installing a radon reduction system in your home. It is important to verify with your contractor that the radon mitigation standards (ASTM E2121 in particular) are properly met to ensure that your radon reduction system will be effective. You also can check with your state radon office to see if there are state requirements that your contractor must meet.

☐ **Radon reduction systems must be clearly labeled. This will avoid accidental changes to the system that could disrupt its function.**

☐ **The exhaust pipes of soil suction systems must vent above the surface of the roof and 10 feet or more above the ground, and must be at least 10 feet away from windows, doors or other openings that could allow radon to reenter the home, if the exhaust pipes do not vent at least 2 feet above these openings.**

☐ **The exhaust fan must not be located in or below a livable area. For instance, it should be installed in unconditioned space.**

☐ **If installing an exhaust fan outside, the contractor must install a fan that meets local building codes for exterior use.**

☐ **Electrical connections of all active radon reduction systems must be installed according to local electrical codes.**

☐ **A warning device must be installed to alert you if an active system stops working properly. Examples of system failure warning devices are: a liquid gauge, a sound alarm, a light indicator, and a dial, or needle display, gauge. The warning device must be placed where it can be seen or heard easily. Your contractor should check that the warning device works. Later on, if your monitor shows that the system is not working properly, call a contractor to have it checked.**

- [ ] A post-mitigation radon test should be done within 30 days of system installation, but no sooner than 24 hours after your system is in operation with the fan on, if it has one. The contractor may perform a post-mitigation test to check his work and the initial effectiveness of the system; however, it is recommend that you also get an independent follow-up radon measurement. Having an independent tester perform the test, or conducting the measurement yourself, will eliminate any potential conflict of interest. To test the system's effectiveness, a two- to seven-day measurement is recommended. Test conditions: windows and doors must be closed 12 hours before and during the test, except for normal entry and exit.

- [ ] Make sure your contractor completely explains your radon reduction system, demonstrates how it operates and explains how to maintain it. Ask for written operating and maintenance instructions and copies of any warranties.

LIVING IN A HOME WITH A RADON REDUCTION SYSTEM

Maintaining Your Radon Reduction System

Similar to a furnace or chimney, radon reduction systems need occasional maintenance. If you have a fan powered (or active) system, you should look at your warning device, usually a manometer, on a regular basis to make sure the system is working correctly. Fans may last for five years or more — manufacturer warranties tend not to exceed five years — and may then need to be repaired or replaced. The cost to replace a fan varies as it is based on labor and materials. Ask qualified mitigators for estimates before work begins. It is a good idea to retest your home at least every two years to be sure radon levels remain low.

Remember, the fan should NEVER be turned off; it must run continuously for the system to work correctly.

The filter in an HRV requires periodic cleaning and should be changed twice a year. Replacement filters for an HRV are easily changed and are priced between $10 and $25. Ask your contractor where filters can be purchased. Also, the vent that brings fresh air in from the outside needs to be inspected for leaves and debris. The ventilator should be checked annually by a heating, ventilating and air conditioning professional to make sure the air flow remains properly balanced. HRVs used for radon control should run all the time.

Remodeling Your Home after Radon Levels Have Been Lowered

If you decide to make major structural changes to your home after you have had a radon reduction system installed, such as converting an unfinished basement area into living space, ask your radon contractor whether these changes could void any warranties. If you are planning to add a new foundation for an addition to your home, ask your radon contractor what measures should be taken to ensure reduced radon levels throughout the home. After you remodel, retest in the lowest lived-in area to make sure the construction did not reduce the effectiveness of the radon reduction system.

BUYING OR SELLING A HOME?

If you are buying or selling a home and need to make decisions about radon, consult EPA's "Home Buyer's and Seller's Guide to Radon." If you are selling a home that has a radon reduction system, inform potential buyers and supply them with information about your system's operation and maintenance. If you are building a new home, consider that it is almost always less expensive to build radon-resistant features into new construction than it is to fix an existing home that has high radon levels. Ask your builder if he or she uses radon-resistant construction features. Your builder can refer to EPA's document "Building Radon Out: A Step-by-Step Guide On How To Build Radon-Resistant Homes," (https://www.epa.gov/radon/publications-about-radon) or your builder can work with a qualified contractor to design and install the proper radon reduction system. To find a qualified contractor contact your state radon office.

All homes should be tested for radon and elevated radon levels should be reduced. **Even new homes built with radon-resistant features should be tested after occupancy to ensure that radon levels are below 4 pCi/L.** If you have a test result of 4 pCi/L or more, you can have a qualified mitigator add a vent fan to an existing passive system to further reduce the radon level in your home.

RADON IN WATER

Most often, the radon in your home's indoor air can come from two sources, the soil or your water supply. Compared to radon entering your home through water, radon entering your home through soil is usually a much larger risk. If you are concerned about radon and you have a private well, consider testing for radon in both air and water. By testing for radon in both air and water, the results could enable you to more completely assess the radon mitigation options best suited to your situation. The devices and procedures for testing your home's water supply are different from those used for measuring radon in air.

The radon in your water supply poses an inhalation risk and a small ingestion risk. Most of your risk from radon in water comes from radon released into the air when water is used for showering and other household purposes. Research has shown that your risk of lung cancer from breathing radon in air is much larger than your risk of stomach cancer from swallowing water with radon in it.

Radon in your home's water in not usually a problem when its source is surface water. A radon in water problem is more likely when its source is ground water, such as a private well or a public water supply system that uses ground water. Some public water systems treat their water to reduce radon levels before it is delivered to your home. If you are concerned that radon may be entering your home through the water and your water comes from a public water supply, contact your water supplier.

If you've tested your private well and have a radon in water problem, it can be easily fixed. Your home's water supply can be treated in one of two ways; point-of-use or point-of-entry. Point-of-entry treatment for the whole home can effectively remove radon from the water before it enters your home's water distribution system. Point-of-entry treatment usually employs either granular activated carbon, or GAC, filters or aeration systems. While GAC filters usually cost less than aeration systems, filters can collect radioactivity and may require a special method of disposal. Both GAC filters and aeration systems have advantages and disadvantages that should be discussed with your state radon office or a water treatment professional. Point-of-use treatment devices remove radon from your water at the tap, but only treat a small portion of the water you use, such as the water you drink. Point-of-use devices are not effective in reducing the risk from breathing radon released into the air from all water used in the home.

For information on radon in water, testing and treatment, and radon in drinking water standards, or for general help, contact your state radon office https://www.epa.gov/radon/find-information-about-local-radon-zones-and-state-contact-information.

RADON REDUCTION OF VARIOUS MITIGATION TECHNIQUES

Technique	Typical Radon Reduction	Comments
Subslab Suction (Subslab depressurization)	50 to 99 percent	Works best if air can move easily in material under slab.
Passive Subslab Suction	30 to 70 percent	May be more effective in cold climates; not as effective as active subslab suction.
draintile Suction	50 to 99 percent	Can work with either partial or complete drain tile loops.
Block-wall Suction	50 to 99 percent	Only in homes with hollow block-walls; requires sealing of major openings.
Sump-Hole Suction	50 to 99 percent	Works best if air moves easily to sump from under the slab.
Submembrane depresserization in a Crawlspace	50 to 99 percent	Less heat loss than natural ventilation in cold winter climates.
natural ventilation in a Crawlspace	0 to 50 percent	Costs variable.
Sealing of Radon Entry Routes	See Comments	Normally only used with other techniques; proper materials and installation required.
House (Basement) Pressurization	50 to 99 percent	Works best with tight basement isolated from outdoors and upper floors.
natural ventilation	Variable/Temporary	Significant heated or cooled air loss; operating costs depend on utility rates and amount of ventilation.
Heat Recovery ventilation (HRv)	Variable/See comments	Limited use; effectiveness limited by radon concentration or the amount of ventilation air available for dilution by the HRV. Best Applied in limited-space areas like basements.
Private well water Systems: Aeration	95 to 99 percent	Generally more efficient than GAC; requires annual cleaning to maintain effectiveness and to prevent contamination; requires venting radon to outdoors.
Private well water Systems: Granular Activated Carbon, or GAC	85 to 95 percent	Less efficient for higher levels than aeration; use for moderate levels, around 5,000 pCi/L or less in water: radioactive radon by-products can build on carbon; may need radiation shield around tank and care in disposal.

Note: Mitigation costs vary due to technique, materials, and the extent of the problem. Typically the cost of radon mitigations are comparable to other common home repairs.

FOR FURTHER INFORMATION

EPA Radon Web site
https://www.epa.gov/radon
EPA's main radon home page. Includes links to publications, hotlines, private proficiency programs and more.

EPA Regional Offices
https://www.epa.gov/radon/find-information-about-local-radon-zones-and-state-contact-information. Check this Web site for a listing of your EPA regional office.

EPA Publications
Most EPA radon publications are available online at https://www.epa.gov/radon/publications-about-radon.

Hotlines

1-800-SOS-RADON (767-7236)
Operated by Kansas State University in partnership with EPA. Order radon test kits by phone.

1-800-426-4791
Safe Drinking Water Hotline, privately operated under contract to EPA. For general information on drinking water, radon in water, testing and treatment and radon drinking water standards.

Proficiency Programs

National Radon Proficiency Program (NRPP)
1-800-269-4174
www.aarst-nrpp.com/wp/

National Radon Safety Board (NRSB)
1-866-329-3474
www.nrsb.org

www.ingramcontent.com/pod-product-compliance
Lightning Source LLC
Chambersburg PA
CBHW081307170526
45165CB00010B/3291

* 9 7 8 1 5 4 8 6 8 3 9 5 5 *